The Open University

Mathematics Foundation Course Unit 28

LINEAR ALGEBRA IV

Prepared by the Mathematics Foundation Course Team

Correspondence Text 28

The Open University Press

Open University courses provide a method of study for independent learners through an integrated teaching system including textual material, radio and television programmes and short residential courses. This text is one of a series that make up the correspondence element of the Mathematics Foundation Course.

The Open University's courses represent a new system of university level education. Much of the teaching material is still in a developmental stage. Courses and course materials are, therefore, kept continually under revision. It is intended to issue regular up-dating notes as and when the need arises, and new editions will be brought out when necessary.

Further information on Open University courses may be obtained from The Admissions Office, The Open University, P.O. Box 48, Bletchley, Buckinghamshire.

The Open University Press
Walton Hall, Bletchley, Bucks

First Published 1971
Copyright © 1971 The Open University

Printed in Great Britain by
J W Arrowsmith Ltd, Bristol 3

SBN 335 01027 X

Contents

Objectives

The principal objective of this unit is to discuss the problem of solving a system of simultaneous linear equations and, in particular, to discuss the accuracy of the results obtained and the efficiency of the method used.

After working through this unit you should be able to:

(i) explain what is meant by the following terms:
error vector,
ill-conditioned system of equations,
nearly singular matrix,
norm of a vector;

(ii) determine the efficiency of the Gauss elimination method or a matrix inversion method, by determining the number of specific arithmetic operations required;

(iii) rearrange a given matrix equation

$$A\underline{x} = \underline{b}$$

into a suitable form which guarantees the convergence of an iterative method for the solution;

(iv) determine whether a given (simple) system of simultaneous equations is ill-conditioned.

Note

Before working through this correspondence text, make sure you have read the general introduction to the mathematics course in the Study Guide, as this explains the philosophy underlying the whole course. You should also be familiar with the section which explains how a text is constructed and the meanings attached to the stars and other symbols in the margin, as this will help you to find your way through the text.

Structural Diagram

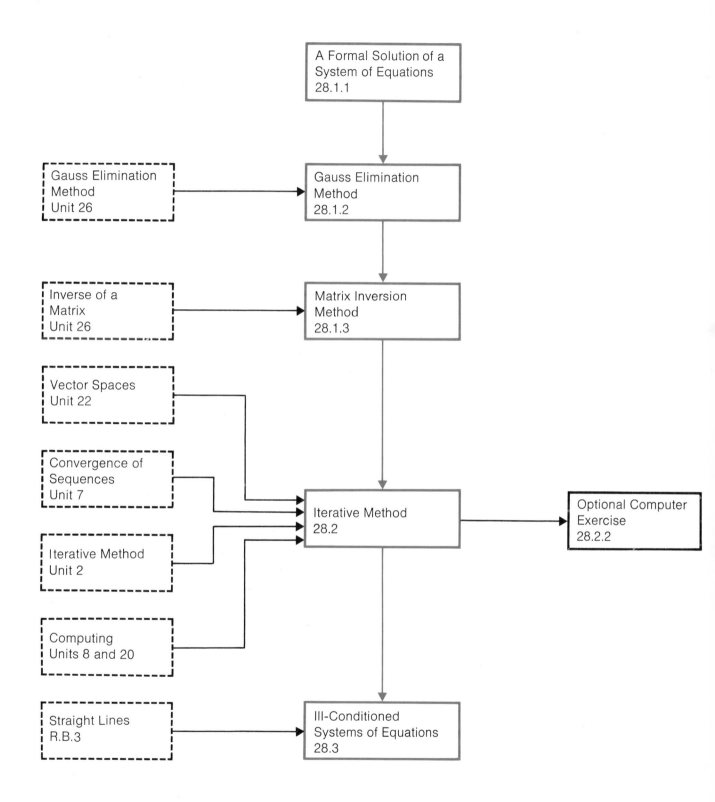

Glossary

Terms which are defined in this text are printed in CAPITALS.

DIRECT METHOD	A DIRECT METHOD of solving a system of linear simultaneous equations is a method which produces an explicit result for the solution.	2
ERROR VECTOR	The ERROR VECTOR for a system of linear equations is *(estimated solution vector)* − *(solution vector)*.	19
ILL-CONDITIONED SYSTEM OF EQUATIONS	An ILL-CONDITIONED SYSTEM OF EQUATIONS is a system for which small changes in the coefficients produce large changes in the elements of the solution set.	37
INDIRECT METHOD	An INDIRECT METHOD in the context of this unit is an iterative method.	16
MATRIX OF COEFFICIENTS	The MATRIX OF COEFFICIENTS of the system of equations	16

$$a_{11}x_1 + a_{12}x_2 + \cdots + a_{1n}x_n = b_1$$
$$a_{21}x_1 + a_{22}x_2 + \cdots + a_{2n}x_n = b_2$$
$$\cdots$$
$$a_{m1}x_1 + a_{m2}x_2 + \cdots + a_{mn}x_n = b_m$$

is

$$\begin{pmatrix} a_{11} & a_{12} & \cdots & a_{1n} \\ a_{21} & a_{22} & \cdots & a_{2n} \\ \cdots & & & \\ a_{m1} & a_{m2} & \cdots & a_{mn} \end{pmatrix}.$$

NEARLY SINGULAR MATRIX	A NEARLY SINGULAR MATRIX is a non-singular matrix for which small changes in the elements would produce a singular matrix.	39
NORM OF A VECTOR	A NORM OF A VECTOR is a numerical measure of the "size" of a vector.	24
SPARSE MATRIX	A SPARSE MATRIX is a matrix in which the elements are predominantly zero.	16

Notation

The symbols are presented in the order in which they appear in the text.

$$(\underline{x}^{(r)} - \underline{X}).$$

$$\|\underline{a}\| = |a_1| + |a_2| + \cdots + |a_n|,$$

where $\underline{a}$ is an $n \times 1$ vector with elements a_i, $i = 1, \ldots, n$.

Bibliography

L. Fox, *An Introduction to Numerical Linear Algebra* (Clarendon Press, 1964).
This book is a very comprehensive survey of the theory and practice of direct and indirect methods of solving simultaneous linear equations and of the associated error analysis. It goes far beyond the scope of this text, but it is referred to at several points for the benefit of the student who wishes to go further into the subject.

A. Ralston, *A First Course in Numerical Analysis* (McGraw-Hill, 1965).
This book discusses many aspects of numerical analysis as well as those involved with the solution of simultaneous linear equations. Chapter 9 on the solution of simultaneous linear equations would be good reading for further study, particularly from the viewpoint of error analysis.

Be prepared: you will find that both books are quite hard reading if your only knowledge of the topic is based on the material in this unit.

L. Fox, *An Introduction to Numerical Linear Algebra* (Clarendon Press, 1964).

28.0 INTRODUCTION

This unit considers in detail just one aspect of the work discussed in *Unit 26, Linear Algebra III*. We are going to look at some practical methods of solving a set of simultaneous linear equations, and some of the difficulties involved. We regard this problem as a vehicle for discussing some of the methods and practices of numerical calculation: it is another step in the numerical analysis aspect of this course, which we began in *Unit 2, Errors and Accuracy*.

We begin by stating the problem precisely.

The matrix equation

$$A\underline{x} = \underline{b},$$

where $\underline{x}$ and $\underline{b}$ are n-element column vectors and A is an $n \times n$ matrix, represents a system of n simultaneous linear equations in n variables (unknowns). If A is a non-singular matrix, that is, its rank is n, then the matrix equation has a *unique* solution. This is the only type of equation we shall deal with in this unit, since one of our aims is to examine how the solution is derived, particularly for large matrices. Thus we shall assume that a unique solution exists; our problem is to determine it.

If you have solved simultaneous equations already, you have probably solved systems of 2 or 3 or perhaps 4 equations. In your working you endeavoured to avoid blunders, and, as a final check, you substituted your answers back into the original equations. You were not interested in comparing different methods of solution to see if some were quicker than others, nor were you concerned with the accuracy of the solution other than in the sense that it satisfied the equation. With systems of several thousand equations, time can be an important factor, and we shall see that there is more to accuracy than correct arithmetical calculations. When you solved your equations, you almost certainly used a *direct* method, i.e. one that leads step by step to the answer, which appears at the last step(s); for instance, a method like the Gauss elimination method which we discussed in *Unit 26*. We shall look at that method again, along with two other direct methods, in section 28.1.

When you solved a small system of equations, you probably never even contemplated an iterative method (i.e. a method in which at each step we obtain another approximation to the solution), since the direct method was so quick and reliable. In larger systems with certain characteristics, iterative methods can be useful. We look at these methods in section 28.2 and we see how they link together the ideas of the iterative methods of solving equations (described in *Unit 2, Errors and Accuracy*) and some of the vector space concepts discussed in the earlier units on linear algebra.

Finally, we come to the problem of accuracy. Frequently, the data we use to set up the equations are inaccurate, and so the eventual result may be in error, not only because of round-off errors which may have built up during the computation, but also from inaccuracies propagated from the very start. In certain special cases this inaccuracy makes the results almost worthless. It is this aspect of the solution of simultaneous equations, called *ill-conditioning*, which we shall look at in section 28.3.

28.1 DIRECT METHODS

28.1.0 Introduction

Mathematicians like to be able to record an explicit solution to an equation. The oft-quoted solution to the quadratic equation

$$ax^2 + bx + c = 0,$$

namely

$$x = \frac{-b \pm \sqrt{b^2 - 4ac}}{2a},$$

is an example of this. The desired variable is to the left of an "equals" sign, and the letters to the right of the "equals" sign can be replaced by numbers from the data in any specific example. The solution consists of the two numbers which map to zero under the function

$$x \longmapsto ax^2 + bx + c \qquad (x \in C).$$

Similarly, we can write down an explicit solution to the matrix equation

$$A\underline{x} = \underline{b},$$

where A is non-singular, in the form

$$\underline{x} = A^{-1}\underline{b},$$

the solution vector being the unique vector which maps to $\underline{b}$ under the mapping

$$\underline{x} \longmapsto A\underline{x} \qquad\qquad (\underline{x} \in R^n).$$

A method which uses such an explicit formula to obtain the answer is called a direct method. Not all direct methods, however, use formulas to calculate the answer. For instance, the Gauss elimination method (described in *Unit 62*) is a direct method, but we do not *use* a formula for the answer, but a step by step procedure (which could be specified by an algorithm) to get from the data to the answer. This is the characteristic of a direct method (as opposed to an iterative method, which we discuss later): it is a method of obtaining the answer to a problem from the data by a step by step procedure, the answer (or an approximation to the answer) being obtained only at the end of the procedure, there being no approximations to the answer en route.

To *calculate* the solution vector to a system of equations using the formula $\underline{x} = A^{-1}\underline{b}$, we must determine A^{-1} and post-multiply by $\underline{b}$. We examine the efficiency of this particular process in section 28.1.3. We shall also compare its efficiency with the efficiency of two other methods: the Gauss elimination method, which will be examined in section 28.1.2, and one other direct method, which we look at in section 28.1.1. Finally, we must emphasize that all three methods of solution are algebraically equivalent, since the solution is unique. This means that if we wrote them out as explicit formulas, we could derive one formula from the other by algebraic manipulation. In this unit we are interested in comparing the computational methods and the lengths of time taken.

28.1.1 An Explicit Method

By appropriate manipulation (for example, eliminating x_2 between the equations, and then solving for x_1), the solution of

$$a_{11}x_1 + a_{12}x_2 = b_1$$

$$a_{21}x_1 + a_{22}x_2 = b_2$$

may be written as

$$x_1 = \frac{b_1 a_{22} - b_2 a_{12}}{a_{11}a_{22} - a_{12}a_{21}}, \qquad x_2 = \frac{a_{11}b_2 - a_{21}b_1}{a_{11}a_{22} - a_{12}a_{21}},$$

provided that

$$a_{11}a_{22} - a_{12}a_{21} \neq 0.$$

(All coefficients $a_{11}, a_{12}, \ldots$ and $b_1, b_2, \ldots$ used in this text are assumed to be real numbers.) In this case the solution we have found is the set of components of the unique vector $\underline{x}$ which maps to $\underline{b}$ under the mapping

$$\underline{x} \longmapsto A\underline{x} \qquad (\underline{x} \in R^2).$$

Thus, at one step, provided the condition is satisfied, we have formally written down the solution to *all* pairs of simultaneous equations in two variables, which have a unique solution.

We can go on to find that for a system of three simultaneous linear equations:

$$a_{11}x_1 + a_{12}x_2 + a_{13}x_3 = b_1$$

$$a_{21}x_1 + a_{22}x_2 + a_{23}x_3 = b_2$$

$$a_{31}x_1 + a_{32}x_2 + a_{33}x_3 = b_3$$

the solution set may be written as

$$x_1 = \frac{b_1(a_{22}a_{33} - a_{23}a_{32}) - b_2(a_{12}a_{33} - a_{13}a_{32}) + b_3(a_{12}a_{23} - a_{13}a_{22})}{a_{11}(a_{22}a_{33} - a_{23}a_{32}) - a_{21}(a_{12}a_{33} - a_{13}a_{32}) + a_{31}(a_{12}a_{23} - a_{13}a_{22})},$$

together with similar expressions for x_2 and x_3 (with the same denominator), provided, of course, that the denominator does not equal zero. (If you have plenty of time to spare, you may like to verify the above expression by solving the system of equations yourself. But don't get too bogged down by the algebraic manipulations.)

Again, we can see that, by simple substitution of numbers for the letters, we can solve *any* system of the above form which has a unique solution. We could, in theory, solve a system of simultaneous linear equations of any order by developing the appropriate formulas. Why then do we not stop here and simply solve simultaneous linear equations this way? The reason is that the expenditure of time and effort involved is too great.

Let us consider the number of multiplications and divisions involved in solving the three simultaneous linear equations. We concentrate on the operations of multiplication and division, since these tend to be more time-consuming than addition and subtraction, both for manual and computer operations.

The expressions for x_1, x_2 and x_3 have the same denominator, so this has to be calculated just once. Each numerator has the same form as the denominator, and there are 3 numerators. Thus we need to calculate 4 expressions of the form:

$$a_{11}(a_{22}a_{33} - a_{23}a_{32}) - a_{21}(a_{12}a_{33} - a_{13}a_{32}) + a_{31}(a_{12}a_{23} - a_{13}a_{22}).$$

Each bracket contains 2 products. Evaluation of the whole expression therefore requires 9 multiplications.

Total number of multiplications is 4×9 $= 36$
Number of divisions $= 3$
Therefore total number of time-consuming operations $= 39$

You may have noticed that three bracketed terms in one numerator (that of x_1 in our example) are the same as the bracketed terms in the denominator. Using this, we could save 6 multiplications and would therefore require only 33 time-consuming operations.

Exercise 1

Exercise 1
(2 minutes)

What is the total number of operations of multiplication and division required to solve a system of two simultaneous linear equations by substitution in the formulas given in the text? ◼

A general expression can be derived for the number of operations of multiplication and division for the solution of n equations in n unknowns, by this method; it is given by the sequence

$$u_2 = 8$$

$$u_n = 2n + n^2 \left(\frac{u_{n-1}}{n-1} - 1 \right), \qquad n > 2.$$

Discussion
* *

For large values of n, the nth term of this sequence, u_n, can be shown to be approximately equal to

$$1.72 \times n \times n!$$

This enables us to produce the table below. The final column gives a rough idea of the time it would take an automatic computer to solve the problem if, for example, each operation were to take one microsecond (1 microsecond $= 10^{-6}$ s; it is denoted by 1 μs).

Number of simultaneous equations	Number of operations	Time taken
2	8	8 μs
3	33	33 μs
4	168	168 μs
10	$\simeq 6.2 \times 10^7$	62 s
100	$\simeq 1.6 \times 10^{160}$	$\simeq 10^{147}$ year
1000	$\simeq 6.9 \times 10^{2570}$	$\simeq 10^{2557}$ year

This shows the tremendous increase in the number of operations and the consequent time required as n increases; a more efficient method is obviously desirable.

Optional Section on Determinants

Optional
Material
*

If you wish, you may omit this section, which is not essential to the development of this unit or the course. It is included because it briefly discusses *determinants*, which have in the past been closely associated with the solution of simultaneous linear equations.

The expression

$$a_{11}a_{22} - a_{12}a_{21}$$

is called the *determinant* of the matrix

$$A_2 = \begin{pmatrix} a_{11} & a_{12} \\ a_{21} & a_{22} \end{pmatrix},$$

and is written as

$$\begin{vmatrix} a_{11} & a_{12} \\ a_{21} & a_{22} \end{vmatrix}$$

or det A_2 or $|A_2|$, so that the solution of two equations in two variables may be written as

$$x_1 = \frac{\begin{vmatrix} b_1 & a_{12} \\ b_2 & a_{22} \end{vmatrix}}{\begin{vmatrix} a_{11} & a_{12} \\ a_{21} & a_{22} \end{vmatrix}} \text{ etc.}$$

Similarly, the expression

$$a_{11}(a_{22}a_{33} - a_{23}a_{32}) - a_{21}(a_{12}a_{33} - a_{13}a_{32}) + a_{31}(a_{12}a_{23} - a_{13}a_{22})$$

is called the *determinant* of the matrix

$$A_3 = \begin{pmatrix} a_{11} & a_{12} & a_{13} \\ a_{21} & a_{22} & a_{23} \\ a_{31} & a_{32} & a_{33} \end{pmatrix},$$

and is written as

$$\begin{vmatrix} a_{11} & a_{12} & a_{13} \\ a_{21} & a_{22} & a_{23} \\ a_{31} & a_{32} & a_{33} \end{vmatrix}$$

or det A_3 or $|A_3|$, so that the solution of three equations in three unknowns may be written as

$$x_1 = \frac{\begin{vmatrix} b_1 & a_{12} & a_{13} \\ b_2 & a_{22} & a_{23} \\ b_3 & a_{32} & a_{33} \end{vmatrix}}{\begin{vmatrix} a_{11} & a_{12} & a_{13} \\ a_{21} & a_{22} & a_{23} \\ a_{31} & a_{32} & a_{33} \end{vmatrix}}.$$

etc.

The determinant of a square matrix is a certain number associated with the matrix. Essentially, its use at this level in mathematics is confined to being a shorthand for expressing the solutions of simultaneous linear equations. Since actual calculations using the formula for the solution, which are equivalent to evaluating the determinants, become much too cumbersome for systems of more than 3 or 4 simultaneous linear equations, we have not made determinants an integral part of the course. The number of equations for which the use of determinants is, so to speak, at its peak performance is only three.

Solution 1

Solution 1

8 ■

28.1.2 The Gauss Elimination Method

28.1.2

Main Text
* *

In *Unit 26*, we introduced the Gauss elimination method of solving systems of equations. The objective of the method is to produce an equivalent set of equations which are easier to solve than the original set. We now ask you to consider this method again and to check how efficient it is in terms of the number of operations it takes to produce the answer, since for large systems of equations the method discussed in section 28.1.1 is clearly unacceptable. (Even if somebody had started a modern computer solving 100 simultaneous equations by that method at the beginning of this century, it would still not have made a dent in the problem: it would not even have performed $10^{-100}\%$ of the calculations.)

Exercise 1

Exercise 1
(3 minutes)

Solve the following set of three equations by the Gauss elimination method with back substitution, in the order shown. Calculate on rough paper and then use the spaces provided to fill in the steps in your solution. The numbers entered should be expressed as decimals, not fractions. Note (in the column at the side) the number of divisions and multiplications that have occurred. What is the total number of these operations for the whole solution?

$$5x_1 + 2x_2 - 2x_3 = 8 \quad (1)$$

$$3x_1 + 6x_2 - 4x_3 = 1 \quad (2)$$

$$2x_1 + 4x_2 - 2x_3 = 1 \quad (3)$$

Step in the Calculation	Number of Multiplications and/or Divisions
Eliminate x_1 from equations (2) and (3).	
Factor for equation (2) is $\boxed{}$	
Factor for equation (3) is $\boxed{}$	
$5x_1 + 2x_2 + (-2) \quad x_3 = \quad 8 \qquad (1)$	
$\boxed{} x_2 + \boxed{} x_3 = \boxed{} \qquad (4)$	
$\boxed{} x_2 + \boxed{} x_3 = \boxed{} \qquad (5)$	
Now eliminate x_2 from equation (5).	
Factor for equation (5) is $\boxed{}$	
$5x_1 + 2x_2 + (-2) \quad x_3 = \quad 8 \qquad (1)$	
$\boxed{} x_2 + \boxed{} x_3 = \boxed{} \qquad (4)$	
$\boxed{} x_3 = \boxed{} \qquad (6)$	
Therefore	
$x_3 = \boxed{}$	
and back-substituting into equation (4) gives	
$x_2 = \boxed{}$	
and from equation (1)	
$x_1 = \boxed{}$	
Total number of divisions and multiplications	

Solution I **Solution 1**

Step in the Calculation	Number of Multiplications and/or Divisions
Eliminate x_1 from equations (2) and (3).	
Factor for equation (2) is $\quad -0.6$	1 division
Factor for equation (3) is $\quad -0.4$	1 division
$5x_1 + 2x_2 + (-2)\quad x_3 = \quad 8 \qquad (1)$	
$4.8 \quad x_2 + \quad -2.8 \quad x_3 = \quad -3.8 \qquad (4)$	3 multiplications
$3.2 \quad x_2 + \quad -1.2 \quad x_3 = \quad -2.2 \qquad (5)$	3 multiplications
Now eliminate x_2 from equation (5).	
Factor for equation (5) is $\quad -0.6\dot{6}$	1 division
$5x_1 + 2x_2 + (-2)\quad x_3 = \quad 8 \qquad (1)$	
$4.8 \quad x_2 + \quad -2.8 \quad x_3 = \quad -3.8 \qquad (4)$	
$0.6\dot{6} \quad x_3 = \quad 0.3\dot{3} \qquad (6)$	2 multiplications
Therefore	
$x_3 = \quad 0.5$	1 division
and back-substituting into equation (4) gives	
$x_2 = \quad -0.5$	1 multiplication and 1 division
and from equation (1)	
$x_1 = \quad 2$	2 multiplications and 1 division
Total number of divisions and multiplications	17

Notice that the number of multiplications and divisions is roughly half the corresponding number for the method considered in section 28.1.1. ■

Next we try to determine the amount of computation required to solve a general system of equations by the Gauss elimination method. We attempt an analysis similar to the one given in the last exercise. We then compare the labour involved in this method with the labour involved in the method of section 28.1.1.

Consider the first two equations of a system of n equations:

$$a_{11}x_1 + a_{12}x_2 + \cdots + a_{1n}x_n = b_1$$

$$a_{21}x_1 + a_{22}x_2 + \cdots + a_{2n}x_n = b_2$$

We assume that a_{11} is non-zero. If it were not, then we could alter the order of the equations until we found a coefficient of x_1 which was non-zero. In hand calculation, such a re-ordering does not involve an important time factor, but in an automatic machine calculation it would, of course, have to be taken into account. To eliminate x_1 from the second equation requires the calculation of the quotient a_{21}/a_{11} and then the formation of the n numbers

$$a_{22} - \frac{a_{21}}{a_{11}}a_{12}, a_{23} - \frac{a_{21}}{a_{11}}a_{13}, \ldots, a_{2n} - \frac{a_{21}}{a_{11}}a_{1n}, b_2 - \frac{a_{21}}{a_{11}}b_1.$$

Thus the formation of the new second equation requires $(n + 1)$ operations of multiplication or division. In the next exercise we ask you to calculate the total number of multiplications and divisions required.

We assume that $a_{22} - \frac{a_{21}}{a_{11}}a_{12}$ is non-zero for the next round, since we shall have to divide by it to produce the next quotient corresponding to $\frac{a_{21}}{a_{11}}$, that is,

$$\frac{a_{32}}{a_{22} - \frac{a_{21}}{a_{11}}a_{12}}.$$

If it were zero, we could alter the order of the equations until we found a coefficient of x_2 which was non-zero. (What would you infer if *all* subsequent coefficients of x_2 were zero?)

In general, in the discussion in this text we shall assume that we can perform the divisions as we come to them.

Exercise 2

Exercise 2
(4 minutes)

Fill in the following table to calculate the number of multiplications and divisions involved in solving a system of n equations in n unknowns by the Gauss elimination method. The formulas derived in *Unit 4, Finite Differences*:

$$S_1(n) = \sum_{r=1}^{n} r = \frac{n(n + 1)}{2}$$

$$S_2(n) = \sum_{r=1}^{n} r^2 = \frac{n(n + 1)(2n + 1)}{6},$$

will be useful.

Step in the Calculation	Number of Multiplications/Divisions
Eliminate x_1 from all equations after the first.	$(n-1)(n+1) = n^2 - 1$
Eliminate x_2 from all equations after the second. $\vdots$	$=$
Eliminate x_{n-1} from all equations after the $(n-1)$th, i.e. the nth equation.	
Total for elimination is	
	$=$
Now carry out the back substitution.	
To determine x_n from an equation such as $\alpha_n x_n = \beta_n$	
To determine x_{n-1} from an equation such as $\gamma_{n-1}x_{n-1} + \gamma_n x_n = \beta_{n-1}$ $\vdots$	
To determine x_1	
Total for back substitution is	
Therefore total number of operations is	

Compare your answer with the answer to Exercise 1. ∎

We are now in a position to compare the explicit solution and the Gauss elimination method from the viewpoint of efficiency. The results are tabulated below.

Number of equations	Formula method No. of operations is $1.72n \times n!$ (for large n)	Gauss elimination method No. of operations is $\dfrac{n^3}{3} + n^2 - \dfrac{n}{3}$
3	33	17
4	168	36
10	$\simeq 6.2 \times 10^7$	430
100	$\simeq 1.6 \times 10^{160}$	$\simeq 3.4 \times 10^5$
1000	$\simeq 6.9 \times 10^{2570}$	$\simeq 3.3 \times 10^8$

This table demonstrates dramatically why the formula method is never used for large systems of equations.

In fact, the formula method is never quicker than the Gauss elimination method. For a quick comparison of the methods for large n, we would say that the number of operations for the Gauss elimination method is approximately $\dfrac{n^3}{3}$, since, when $n > 100$ say, the rest of the terms affect the number of operations by less than three per cent.

Solution 2 **Solution 2**

Step in the Calculation	Number of Multiplications/Divisions
Eliminate x_1 from all equations after the first.	$(n-1)(n+1) = n^2 - 1$
Eliminate x_2 from all equations after the second. $\vdots$	$(n-2)n = (n-1)^2 - 1$
Eliminate x_{n-1} from all equations after the $(n-1)$th, i.e. the nth equation.	$2^2 - 1$
Total for elimination is	$\left(\displaystyle\sum_{r=2}^{n} r^2\right) - (n-1)$
	$= \left(\dfrac{n(n+1)(2n+1)}{6} - 1\right) - (n-1)$
Now carry out the back substitution.	
To determine x_n from an equation such as $\alpha_n x_n = \beta_n$	1
To determine x_{n-1} from an equation such as $\gamma_{n-1} x_{n-1} + \gamma_n x_n = \beta_{n-1}$ $\vdots$	2
To determine x_1	n
Total for back substitution is	$\displaystyle\sum_{r=1}^{n} r = \dfrac{n(n+1)}{2}$
Therefore total number of operations is	$\dfrac{n^3}{3} + n^2 - \dfrac{n}{3}$

$\blacksquare$

28.1.3 A Matrix Inversion Method

In *Unit 26* we described a numerical method for finding the inverse of an $n \times n$ matrix; the method is very similar to the Gauss elimination method. We pointed out that if we had to solve several systems of equations of the form

$$A\underline{x} = \underline{b},$$

in which the matrix A was the same in each case, and only the matrix $\underline{b}$ changed, then there might be some value in computing A^{-1} and calculating the solution from the formula

$$\underline{x} = A^{-1}\underline{b}$$

for each case.

We shall now look at the number of calculations involved and see just when this method is of "some value".

Just to remind you of the method, we describe the essential steps. Suppose we want to invert the matrix

$$\begin{pmatrix} 2 & 1 & -1 \\ 6 & -1 & -9 \\ 4 & 3 & 1 \end{pmatrix}.$$

We write

$$\begin{array}{ccc|ccc} 2 & 1 & -1 & 1 & 0 & 0 \\ 6 & -1 & -9 & 0 & 1 & 0 \\ 4 & 3 & 1 & 0 & 0 & 1 \end{array}$$

and use elementary row operations to turn this array into

$$\begin{array}{ccc|ccc} 1 & 0 & 0 & \times & \times & \times \\ 0 & 1 & 0 & \times & \times & \times \\ 0 & 0 & 1 & \times & \times & \times \end{array}$$

where the crosses represent the elements of the inverse matrix. We have ignored the row–sum check: this would be involved both here and in the Gauss elimination method, so for a comparison between the two methods we can ignore it.

We begin by dividing the first row of our array of numbers by 2 in order to obtain a 1 in the top left-hand corner. This involves 3 divisions. We obtain

$$\begin{array}{ccc|ccc} 1 & \frac{1}{2} & -\frac{1}{2} & \frac{1}{2} & 0 & 0 \\ 6 & -1 & -9 & 0 & 1 & 0 \\ 4 & 3 & 1 & 0 & 0 & 1 \end{array}$$

Next we subtract 6 times the first row from the second and 4 times the first row from the third. This involves 2×3 multiplications. We obtain

$$\begin{array}{ccc|ccc} 1 & \frac{1}{2} & -\frac{1}{2} & \frac{1}{2} & 0 & 0 \\ 0 & -4 & -6 & -3 & 1 & 0 \\ 0 & 1 & 3 & -2 & 0 & 1 \end{array}$$

Thus to get the first column into the required form we have performed 9 time-consuming operations.

In general, to compute the inverse of an $n \times n$ matrix, we form the corresponding $n \times 2n$ matrix by "joining" on the $n \times n$ unit matrix.

To produce

$$\begin{pmatrix} 1 \\ 0 \\ \vdots \\ 0 \end{pmatrix}$$

in the first column requires n divisions to produce the 1 and $(n-1) \times n$ operations to produce the 0's. That means we require $n + (n-1) \times n = n^2$ time-consuming operations to derive the first column. Performing the manipulations to produce each subsequent column of an $n \times n$ unit matrix also requires n^2 operations. Since there are n columns, the total number of operations required is n^3.

Having found the inverse matrix, we then have to calculate $A^{-1}\underline{b}$, which involves a further n^2 multiplications.

Thus the total number of operations to obtain a solution is

$$n^3 + n^2,$$

compared with

$$\frac{n^3}{3} + n^2 - \frac{n}{3}$$

for the Gauss elimination method.

Number of equations	Gauss elimination No. of operations	Inverse method No. of operations
3	17	36
4	36	80
10	430	1100
100	$\simeq 3.4 \times 10^5$	$\simeq 3 \times 3.4 \times 10^5$
1000	$\simeq 3.3 \times 10^8$	$\simeq 3 \times 3.3 \times 10^8$

For large n, the n^3 term is much larger than the rest, so that the inverse method is roughly three times as long as the Gauss elimination method. Thus, when there are *three or more* systems of simultaneous equations to solve with the *same A* but *different $\underline{b}$*'s, it is usually quicker to find A^{-1} first, and use it for all the different vectors $\underline{b}$, rather than calculate each solution by the Gauss elimination method.

Exercise 1

How many multiplications are required to multiply together two $n \times n$ matrices? ■

Exercise 1
(3 minutes)

28.1.4 Summary

In this section we have measured the efficiency, in terms of time for computation, of three direct methods of solving systems of n equations (in n unknowns) by investigating the number of operations of multiplication and division required for each. For large n, we found the following results.

	Approximate number of operations	Usefulness for large n
Explicit method of section 28.1.1.	$1.72n \times n!$	Of no use whatsoever.
Gauss elimination method.	$\dfrac{n^3}{3}$	Widely used when there are no more than 3 systems with the same left-hand side.
A method using the inverse of a matrix.	n^3	Widely used when there are more than 3 systems with the same left-hand side.

The methods considered in section 28.1 are basic methods. There are many refinements to them for particular problems, some of which are listed, for example, in Fox, *Numerical Linear Algebra*, Chapters 3 and 4 (see Bibliography).

$$\begin{pmatrix} \boxed{\text{n terms}} \\ \vdots \end{pmatrix} \begin{pmatrix} \boxed{\text{n terms}} \cdots \cdots \\ \end{pmatrix}$$

Each element of the product matrix requires n multiplications.

There are n^2 terms in the product matrix. Therefore the number of multiplications is n^3. ■

28.2 ITERATIVE OR INDIRECT METHODS

28.2.0 Introduction

Theoretically, given enough time and ignoring round-off errors which may occur in the computation, it is possible to solve exactly a matrix equation

$$A\underline{x} = \underline{b},$$

with exact elements a_{ij} and b_i, by one of the direct methods described in section 28.1. It may therefore appear to be pointless to consider any other method, such as an iterative method in which we make a guess at the solution and then refine it. However, iterative methods are important; we discuss them for the following reasons. First, it often happens that, when there is a large system of equations to be solved, a large number of the elements of the matrix of coefficients are zero; for example,

$$\begin{pmatrix} \times & \times & 0 & 0 & \times & 0 \\ \times & \times & 0 & \times & 0 & 0 \\ 0 & \times & \times & 0 & 0 & 0 \\ \times & 0 & 0 & \times & 0 & \times \\ \times & 0 & 0 & 0 & \times & 0 \\ 0 & 0 & \times & 0 & 0 & \times \end{pmatrix},$$

where the $\times$'s represent non-zero numbers. Such matrices are called sparse matrices (for obvious reasons). An iterative method, by taking account of the zeros from the outset, can sometimes give the result much more quickly than a direct method. Of course, if the pattern of zeros were convenient, for example, if we started with

$$\begin{pmatrix} \diagdown & \text{x's} \\ \text{0's} & \diagdown \end{pmatrix}$$

then the back-substitution in the Gauss elimination method would be available straight away. The point is that the iterative method does not depend on a convenient pattern. Secondly, the iterative method, if it converges, improves the accuracy of the approximate solution at each stage. As such it can be used, after a first crude approximation by a direct method, to improve the accuracy of the solution. There may be other advantages in terms of time and economy in the use of space in a digital computer, but these would need closer analysis. In terms of automatic computation, iterative methods often have the advantage that various stages in the iteration repeat the *same* relatively simple process, which is ideal for economic and efficient programming. The iterative method is also of interest from a purely mathematical viewpoint; it shows how the ideas of the mapping of numerical error intervals, introduced in *Unit 2, Errors and Accuracy*, can be extended to the discussion of vectors.

28.2.1   Some Methods of Iteration

We recapitulate briefly on the iterative method for solving equations defined on the set of real numbers, which we discussed in *Unit 2, Errors and Accuracy*. To solve the equation

$$x^3 - 5x + 3 = 0,$$

we tried a rearrangement such as

$$x = \tfrac{1}{5}(x^3 + 3),$$

which picks out the fixed or *invariant* elements of the domain of the function

$$x \longmapsto \tfrac{1}{5}(x^3 + 3) \qquad (x \in R),$$

i.e. those elements of the domain which are equal to their images in the codomain. A first guess, x_0, at the solution gave a new number

$$x_1 = \tfrac{1}{5}(x_0^3 + 3),$$

which we hoped was nearer to the exact solution, which we shall call X. Subsequent steps formed the sequence whose elements x_r were given by

$$x_r = \tfrac{1}{5}(x_{r-1}^3 + 3),$$

which converged to X, provided that the magnitude of a *scale factor*, determined from the function

$$x \longmapsto \tfrac{1}{5}(x^3 + 3) \qquad (x \in R),$$

was less than unity.

In this unit, the equation defined on the set of real numbers is replaced by the matrix equation

$$A\underline{x} - \underline{b} = \underline{0}.$$

The solution, instead of being a number, is now a solution vector. The mapping, which we shall derive from some rearrangement of the matrix equation, such as

$$\underline{x} = G\underline{x} + H\underline{b},$$

now maps an *n*-dimensional vector space to an *n*-dimensional vector space. We look for the invariant vector, i.e. the vector which remains unchanged under the mapping. We shall search for an algorithm giving a sequence of vectors which converges to the solution vector. By analogy with our previous experience of iterative methods, we shall also look for some number, corresponding to the scale factor, which will give us a hint as to whether the sequence under investigation is convergent. A considerable amount of the analysis involved is beyond the scope of this course, but we can get a hint of what it is like by looking at an elementary example. We shall use an illustration of two simultaneous equations in two unknowns although, customarily, the method would be used only on much larger systems.

We shall examine the system of equations

$$x_1 + 4x_2 = 6$$
$$2x_1 + x_2 = 5$$

The actual solution can easily be found to be

$$x_1 = 2, \quad x_2 = 1;$$

that is, the solution vector is

$$\begin{pmatrix} 2 \\ 1 \end{pmatrix}.$$

We postpone the complete matrix approach for a moment and simply rearrange the equations in a couple of ways. Firstly, we solve the first equation for x_2 in terms of x_1 and the second for x_1 in terms of x_2, obtaining

$$x_2 = 1.5 - 0.25x_1$$
$$x_1 = 2.5 - 0.5x_2$$

We now develop the sequences $x_1^{(r)}$, $x_2^{(r)}$ where successive elements* are defined by

$$x_2^{(r+1)} = 1.5 - 0.25x_1^{(r)}$$
$$x_1^{(r+1)} = 2.5 - 0.5x_2^{(r)}$$

Let the first guess at the solution vector,

$$\begin{pmatrix} x_1^{(0)} \\ x_2^{(0)} \end{pmatrix},$$

be

$$\begin{pmatrix} 0 \\ 0 \end{pmatrix}.$$

We then get

$$x_2^{(1)} = 1.5 - 0.25 \times 0 = 1.5$$
$$x_1^{(1)} = 2.5 - 0.5 \times 0 = 2.5$$

and the sequence of estimated solution vectors up to the fifth term is

$$\begin{pmatrix} 0 \\ 0 \end{pmatrix}, \begin{pmatrix} 2.5 \\ 1.5 \end{pmatrix}, \begin{pmatrix} 1.75 \\ 0.88 \end{pmatrix}, \begin{pmatrix} 2.06 \\ 1.06 \end{pmatrix}, \begin{pmatrix} 1.97 \\ 0.98 \end{pmatrix}.$$

Intuitively, it seems that this sequence is converging to the solution vector $\begin{pmatrix} 2 \\ 1 \end{pmatrix}$.

Exercise 1

Exercise 1
(3 minutes)

Use the rearrangement

$$x_1 = 6 - 4x_2$$
$$x_2 = 5 - 2x_1,$$

starting with the vector $\begin{pmatrix} 0 \\ 0 \end{pmatrix}$, to obtain a sequence of five estimated solution vectors for the system of equations in the text. Does it appear to converge? ∎

---

* In previous iterative processes we have used the subscript r, e.g. x_r, to indicate the rth estimate of the solution. Now, to avoid confusion with the subscripts already in use and with the normal use of a superscript as an index, we use a superscript in brackets, e.g. $x^{(r)}$.

Discussion

The last exercise indicates, as we might expect from our previous experience, that some rearrangements "work" and some do not. What we would like to have is some means of testing a rearrangement, corresponding to the scale factor test discussed in *Unit 2*. For example, we might like to test

$$x_1 = 3 + 0.5x_1 - 2x_2$$
$$x_2 = 2.5 - x_1 + 0.5x_2,$$

which is more complicated than either of the previous two rearrangements.

We shall start our search for a "scale factor" in the context of simultaneous equations by examining the iterative process in more general terms. All our rearrangements have the iterative form

$$\underline{x}^{(r+1)} = G\underline{x}^{(r)} + H\underline{b}$$

where G and H are square matrices. For instance, in the rearrangement for which we calculated the sequence which appeared to converge, we had

$$x_1^{(r+1)} = 2.5 - 0.5x_2^{(r)}$$
$$x_2^{(r+1)} = 1.5 - 0.25x_1^{(r)}$$

This can be written in matrix form as

$$\begin{pmatrix} x_1^{(r+1)} \\ x_2^{(r+1)} \end{pmatrix} = \begin{pmatrix} 0 & -0.5 \\ -0.25 & 0 \end{pmatrix} \begin{pmatrix} x_1^{(r)} \\ x_2^{(r)} \end{pmatrix} + \begin{pmatrix} 0 & 0.5 \\ 0.25 & 0 \end{pmatrix} \begin{pmatrix} 6 \\ 5 \end{pmatrix}.$$

So in this case,

$$G = \begin{pmatrix} 0 & -0.5 \\ -0.25 & 0 \end{pmatrix} \quad \text{and} \quad H = \begin{pmatrix} 0 & 0.5 \\ 0.25 & 0 \end{pmatrix}.$$

We shall now attempt to explain what we mean by convergence of the sequence of estimated solution vectors. When we are solving an equation in R, we can work in terms of numerical errors in a single number and try to make this as small as possible. We try to find a rearrangement of the equation, and hence a function, which will make the error interval in which the true solution lies as small as possible. Now that we are working in terms of vectors, it seems natural to examine the behaviour of an *error vector*. We define the error vector for the rth iteration to be

Definition 1
* * *

$$\underline{e}^{(r)} = \underline{x}^{(r)} - \underline{X},$$

where $\underline{x}^{(r)}$ is the rth approximation to the exact solution vector $\underline{X}$. When the sequence converges to $\underline{X}$, the error vector approaches the zero vector. We require the error vector $\underline{e}^{(r)}$ to get "smaller", in some sense yet to be defined, as r increases.

The solution vector $\underline{X}$ must satisfy the equation

$$\underline{X} = G\underline{X} + H\underline{b},$$

since this is merely a rearrangement of the equation

$$A\underline{X} - \underline{b} = \underline{0},$$

just as

$$x = \tfrac{1}{5}(x^3 + 3)$$

is a rearrangement of the equation

$$x^3 - 5x + 3 = 0.$$

(*continued on page 20*)

Solution 1

Solution 1

$$\begin{pmatrix} 0 \\ 0 \end{pmatrix}, \quad \begin{pmatrix} 6 \\ 5 \end{pmatrix}, \quad \begin{pmatrix} -14 \\ -7 \end{pmatrix}, \quad \begin{pmatrix} 34 \\ 33 \end{pmatrix}, \quad \begin{pmatrix} -126 \\ -63 \end{pmatrix}$$

It does not appear to converge. ■

(continued from page 19)

From the two equations

$$\underline{x}^{(r+1)} = G\underline{x}^{(r)} + H\underline{b}$$

$$\underline{X} = G\underline{X} + H\underline{b},$$

we can obtain a relationship between successive error vectors. We have

$$\underline{x}^{(r+1)} - \underline{X} = G\underline{x}^{(r)} - G\underline{X}$$

$$= G(\underline{x}^{(r)} - \underline{X}),$$

i.e.

$$\underline{e}^{(r+1)} = G\underline{e}^{(r)}.$$

In the next section we shall use this result and establish a measure for convergence. At first glance, it may seem that it is easy enough to judge whether $\underline{e}^{(r+1)}$ is "smaller" than $\underline{e}^{(r)}$; for instance, we can say that the individual elements of $\underline{e}^{(r+1)}$ must all be smaller in absolute magnitude than the corresponding elements of $\underline{e}^{(r)}$. But, on reflection, this is seen to be unsatisfactory. What if *nearly all* the elements of $\underline{e}^{(r+1)}$ are smaller than the corresponding elements of $\underline{e}^{(r)}$, but a few are just a little bigger? Looking at the individual elements will not do: we must find a *single number* as a measure.

Exercise 2

Two other rearrangements of our original equations

$$x_1 + 4x_2 = 6$$

$$2x_1 + x_2 = 5$$

are

(i) $x_1 = 6 - 4x_2$

  $x_2 = 5 - 2x_1$

(ii) $x_1 = 3 + 0.5x_1 - 2x_2$

  $x_2 = 2.5 - x_1 + 0.5x_2$

Write down the corresponding iteration formulas in the matrix form

$$\underline{x}^{(r+1)} = G\underline{x}^{(r)} + H\underline{b},$$

and calculate the first five error vectors in each case from the equation

$$\underline{e}^{(r+1)} = G\underline{e}^{(r)},$$

starting with $\underline{e}^{(0)} = \begin{pmatrix} \alpha \\ \beta \end{pmatrix}$. ■

28.2.2 "Measuring" a Vector

To study the convergence of the iterative method we must have some measure by which we can test whether the error vector is getting "smaller" at successive iterations. We can invent a measure in any way we like, and convergence may well depend on the measure we choose. So we must give some thought to what sort of measure is reasonable.

In the first place, we want our measure to be a *single number* because it is easy to compare numbers and to decide whether the sequence of such numbers, obtained from successive error vectors, is convergent to the number associated with the zero error vector. This means that, in the general case, we want to define a function which maps R^n to R.

We begin by considering some unsatisfactory measures, so that we can get a clearer idea of the conditions which a suitable "measure" must satisfy.

Let $\underline{a} = \begin{pmatrix} a_1 \\ a_2 \\ \vdots \\ a_n \end{pmatrix}$ and $\underline{0} = \begin{pmatrix} 0 \\ 0 \\ \vdots \\ 0 \end{pmatrix}$, the zero error vector.

Let us define the function

$$\underline{a} \longmapsto a_1 \qquad (\underline{a} \in R^n),$$

so that a_1 is a "measure" of $\underline{a}$.

This measure is clearly unsatisfactory, for, although the sequence of a_1's obtained from successive error vectors may converge to zero, this fact does not tell us what is happening to the remaining elements in the error vectors.

Consider the function

$$\underline{a} \longmapsto \sum_{i=1}^{n} a_i \qquad (\underline{a} \in R^n).$$

Then,

$$\underline{0} \longmapsto 0,$$

and we want to see if the sequence of "measures" obtained from successive error vectors converges to zero.
But again this is unsatisfactory, because, for instance, if n is even and the error vectors are

$$\underline{e}^{(r)} = \begin{pmatrix} 1 \\ -1 \\ 1 \\ \vdots \\ -1 \end{pmatrix}$$

for all $r > N$, where N is some positive integer, then $\underline{e}^{(r)} \longmapsto 0$, but the error in each element of the estimated solution to our system is not negligible.
The first measure is unsatisfactory because it does not take account of all the elements in the error vector. The second measure is unsatisfactory because, although it does take account of all the elements of the vector, we can get a misleading result if some of the elements are positive and some are negative.

(*continued on page 22*)

Solution 28.2.1.2

(i) $\underline{x}^{(r+1)} = \begin{pmatrix} 0 & -4 \\ -2 & 0 \end{pmatrix} \underline{x}^{(r)} + \begin{pmatrix} 1 & 0 \\ 0 & 1 \end{pmatrix} \underline{b}$

The first five error vectors are

$$\begin{pmatrix} \alpha \\ \beta \end{pmatrix}, \begin{pmatrix} -4\beta \\ -2\alpha \end{pmatrix}, \begin{pmatrix} 8\alpha \\ 8\beta \end{pmatrix}, \begin{pmatrix} -32\beta \\ -16\alpha \end{pmatrix}, \begin{pmatrix} 64\alpha \\ 64\beta \end{pmatrix}.$$

The error vectors are getting bigger by any reasonable "measure", which suggests non-convergence; this agrees with our conclusions in Exercise 1.

(ii) $\underline{x}^{(r+1)} = \begin{pmatrix} 0.5 & -2 \\ -1 & 0.5 \end{pmatrix} \underline{x}^{(r)} + \begin{pmatrix} 0.5 & 0 \\ 0 & 0.5 \end{pmatrix} \underline{b}$

The first five error vectors are

$$\begin{pmatrix} \alpha \\ \beta \end{pmatrix}, \begin{pmatrix} 0.5\alpha - 2\beta \\ -\alpha + 0.5\beta \end{pmatrix}, \begin{pmatrix} 2.25\alpha - 2\beta \\ -\alpha + 2.25\beta \end{pmatrix}, \begin{pmatrix} 3.125\alpha - 5.5\beta \\ -2.75\alpha + 3.125\beta \end{pmatrix},$$

$$\begin{pmatrix} 7.0625\alpha - 9\beta \\ -4.5\alpha + 7.0625\beta \end{pmatrix}.$$

This time it is not so easy to see what is going on. For instance, if we choose $\alpha = \beta = 1$, we get the following error vectors:

$$\begin{pmatrix} 1 \\ 1 \end{pmatrix}, \begin{pmatrix} -1.5 \\ -0.5 \end{pmatrix}, \begin{pmatrix} 0.25 \\ 1.25 \end{pmatrix}, \begin{pmatrix} -2.375 \\ 0.375 \end{pmatrix}, \begin{pmatrix} -1.9375 \\ 2.5625 \end{pmatrix}.$$

Again, we have a sequence of vectors which does not look very hopeful. ∎

(continued from page 21)

We can improve on the second measure in two fairly obvious ways: we define the functions

$$m_1 : \underline{a} \longmapsto \left(\sum_{i=1}^{n} a_i^2 \right)^{1/2} \qquad (\underline{a} \in R^n)$$

and

$$m_2 : \underline{a} \longmapsto \sum_{i=1}^{n} |a_i| \qquad (\underline{a} \in R^n).$$

The measure defined by m_1 is the more obvious for two reasons: the measure is the same as the standard deviation of the a_i from zero; also, in two or three dimensions, this measure can be interpreted as the length or modulus of the corresponding geometric vector.

We shall look at each of these measures briefly to illustrate their use, restricting ourselves to the geometric interpretation in two or three dimensions.

We look again at the two-dimensional example we discussed in the previous section, and, in particular, at the rearrangement which led to the equation

$$\underline{e}^{(r+1)} = G\underline{e}^{(r)}$$

where

$$G = \begin{pmatrix} 0 & -0.5 \\ -0.25 & 0 \end{pmatrix}.$$

If we write

$$\underline{e}^{(r)} = \begin{pmatrix} e_1^{(r)} \\ e_2^{(r)} \end{pmatrix},$$

then

$$m_1 : \underline{e}^{(r)} \longmapsto \sqrt{(e_1^{(r)})^2 + (e_2^{(r)})^2}.$$

If we represent the error vector by an arrow from the origin of a set of Cartesian co-ordinates to the point with co-ordinates $(e_1^{(r)}, e_2^{(r)})$, then $m_1(\underline{e}^{(r)})$ is the length of the arrow. Now, we have

$$\begin{pmatrix} e_1^{(r+1)} \\ e_2^{(r+1)} \end{pmatrix} = \begin{pmatrix} 0 & -0.5 \\ -0.25 & 0 \end{pmatrix} \begin{pmatrix} e_1^{(r)} \\ e_2^{(r)} \end{pmatrix},$$

that is,

$$e_1^{(r+1)} = -0.5\, e_2^{(r)}$$
$$e_2^{(r+1)} = -0.25\, e_1^{(r)}.$$

Let us suppose that the initial error vector $\underline{e}^{(0)} = \begin{pmatrix} \alpha \\ \beta \end{pmatrix}$; then we can draw an arrow from the origin to represent the error vector. The successive error vectors are then represented as follows.

r	$e_1^{(r)}$	$e_2^{(r)}$
0	α	β
1	-0.5β	-0.25α
2	0.125α	0.125β

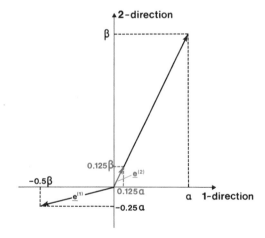

In this particular example, the error vector has one of two directions and alternates between them, so that, after two steps in the iteration, the error vector is again pointing in the same direction, but its length has been decreased by a factor of 8. By taking a sufficient number of steps we can make the length of the error vector as small as we please, i.e. we can get the pointed end of the arrow as close to the origin as we please. This suggests that, with this measure of an error vector, the iteration method will converge to a solution whatever error $\begin{pmatrix} \alpha \\ \beta \end{pmatrix}$ there is in our initial guess which starts the iteration.

We now consider the geometrical interpretation of the second measure, defined by:

$$m_2(\underline{a}) = \sum_{i=1}^{n} |a_i|.$$

With the notation as before, in two dimensions we have

$$m_2(\underline{e}^{(r)}) = |e_1^{(r)}| + |e_2^{(r)}|,$$

which is represented as the sum of the two *lengths* marked in red on the diagram.

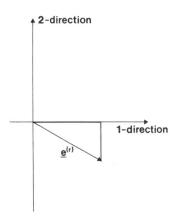

In our example, we see that the sum of the moduli for successive error vectors gets smaller and smaller, so that once again we have an intuitive concept of the error vector converging to the zero vector.

Having gained an intuitive impression of what we mean by convergence for vectors, we shall now be more precise.

A measure of the kind we have been discussing is called a *norm*. A norm is a function with certain special properties which maps the elements of a vector space V to R. The image of a vector $\underline{v}$ under this function is denoted by $\|\underline{v}\|$ (read as "the norm of $\underline{v}$").

Definition 1
* * *

Notation 1
* * *

A norm is defined to have the following properties:

(i) $\|\underline{v}\| \geq 0$, for all $\underline{v} \in V$;
(ii) $\|\underline{v}\| = 0$ if and only if $\underline{v} = \underline{0}$;
(iii) $\|\underline{v}_1 + \underline{v}_2\| \leq \|\underline{v}_1\| + \|\underline{v}_2\|$, for all $\underline{v}_1, \underline{v}_2 \in V$;
(iv) $\|\alpha \underline{v}_1\| = |\alpha| \, \|\underline{v}_1\|$, where α is any real number.

We now define convergence in a vector space as follows. Let m be a norm on the vector space, and let

$$\underline{x}^{(1)}, \underline{x}^{(2)}, \ldots$$

be an infinite sequence of vectors. This sequence is said to have the limit $\underline{X}$ if the sequence of norms of the error vectors:

$$\|\underline{x}^{(1)} - \underline{X}\|, \|\underline{x}^{(2)} - \underline{X}\|, \ldots$$

has limit 0. So we have defined convergence in a vector space in terms of the more familiar notion of convergence in R. It is not our intention to study norms further, so you may love them or leave them. An obvious next step would be to see which of our results for convergence in R can be extended to convergence in a vector space, but we leave this to a later course. We have shown that we can extend the idea of convergence in an apparently satisfactory way, and that is sufficient for now. The next exercise asks you to prove one simple result for the norm m_2 introduced above.

In subsequent work we shall assume that the norm is m_2, since this proves to be more convenient in practice than m_1.

Exercise 1

Consider the norm m_2 in R^n, and show that if the limit of the sequence

$$\underline{x}^{(1)}, \underline{x}^{(2)}, \ldots, \underline{x}^{(k)}, \ldots \text{ is } \underline{X},$$

then the limit of the sequence

$$x_1^{(1)}, x_1^{(2)}, x_1^{(3)}, \ldots, x_1^{(k)}, \ldots$$

is X_1, where

$$\underline{x}^{(k)} = \begin{pmatrix} x_1^{(k)} \\ x_2^{(k)} \\ \vdots \\ x_n^{(k)} \end{pmatrix}, \qquad \underline{X} = \begin{pmatrix} X_1 \\ X_2 \\ \vdots \\ X_n \end{pmatrix}.$$

Notice that this result tells us that, if a sequence of vectors converges, then the sequences of corresponding elements in the vectors themselves converge to the corresponding elements in the limit vector. ∎

Exercise 1
(5 minutes)

Exercise 2

This exercise is intended to give some geometric impression of m_2 in the two-dimensional case. We shall refer to it in the subsequent text.

Let $\underline{e} = \begin{pmatrix} e_1 \\ e_2 \end{pmatrix}$ be an error vector.

(i) Draw the graph of the solution set of

$$m_2(\underline{e}) = |e_1| + |e_2| = 2$$

(i.e. $|x| + |y| = 2$ in Cartesian co-ordinates).
(HINT: Consider the four quadrants of the plane separately.)
What does this imply about the arrow, starting at the origin, representing the error vector?
(ii) What region of the plane represents the solution set of

$$|e_1| + |e_2| \leqslant 2?$$

(i.e. $|x| + |y| \leqslant 2$). ∎

Exercise 2
(4 minutes)

Given a sequence of error vectors, we now have the means to test it for convergence using the norm m_2 in R^n. But this is not ideal, because we first have to obtain the sequence before we can test it. We know that the sequence is obtained from

$$\underline{e}^{(r+1)} = G\underline{e}^{(r)}$$

if our original equation

$$A\underline{x} = \underline{b}$$

was arranged in the form

$$\underline{x}^{(r+1)} = G\underline{x}^{(r)} + H\underline{b}.$$

In other words, the sequence of error vectors is determined by the rearrangement; in particular, it is determined by the matrix G. So we must now determine what conditions we can impose on G (i.e. on the rearrangement) so that the resulting sequence of error vectors converges to $\underline{0}$.

Discussion

(*continued on page 27*)

Solution 1

Solution 1

Since the limit of $\underline{x}^{(1)}, \underline{x}^{(2)}, \ldots$ is $\underline{X}$, we know that the sequence of real numbers

$$\|\underline{x}^{(1)} - \underline{X}\|, \|\underline{x}^{(2)} - \underline{X}\|, \ldots$$

converges to zero.

Therefore, using our definition of the norm m_2, we know that the limit of the sequence whose kth term is

$$|x_1^{(k)} - X_1| + |x_2^{(k)} - X_2| + \cdots + |x_n^{(k)} - X_n|,$$

is zero.

We now need an obvious but unproved result, that is, if $\underline{u}$ and $\underline{v}$ are sequences all of whose terms are positive, and

$$\lim (\underline{u} + \underline{v}) = 0,$$

then both $\underline{u}$ and $\underline{v}$ converge to zero. (You might like to revise your knowledge of *Unit 7, Sequences and Limits I* and prove this result.)

We can split our given sequence into n sequences, one of which is

$$|x_1^{(1)} - X_1|, |x_1^{(2)} - X_1|, \ldots$$

Since all the terms in the expression

$$|x_1^{(k)} - X_1| + |x_2^{(k)} - X_2| + \cdots + |x_n^{(k)} - X_n|$$

are positive, we can apply the result quoted above. Instead of two sequences, we have n sequences, one of which is derived from the first term in the bracket above, i.e.

$$|x_1^{(1)} - X_1|, |x_1^{(2)} - X_1|, \ldots$$

and this, therefore, converges to zero, i.e. the sequence

$$x_1^{(1)}, x_1^{(2)}, \ldots$$

converges to X_1. ■

Solution 2

Solution 2

(i) In the first quadrant of the plane, $x \geqslant 0$, $y \geqslant 0$ and the solution set is

$$x + y = 2, \quad \text{i.e. } y = 2 - x.$$

In the second quadrant of the plane, $x \leqslant 0$, $y \geqslant 0$, so that the solution set is

$$-x + y = 2, \quad \text{i.e. } y = 2 + x,$$

and so on.

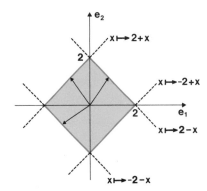

The graph of the solution set is, therefore, the square shown in red on the diagram. The head of the arrow lies *on* the square.

(ii) The region representing the solution set is the interior of the red square. ∎

(continued from page 25)

Now we have met this sort of problem before. In *Unit 2, Errors and Accuracy*, we solved equations in R by rearranging them to get a convergent iterative sequence. We found that there was a criterion associated with the rearrangement which determined convergence: it was a condition on the *scale factor*. Let us just look at this again.

Suppose we want to solve the equation

$$f(x) = 0,$$

where f is a real function, and we have a rearrangement

$$x = F(x)$$

of this equation, leading to the iteration formula

$$x^{(r+1)} = F(x^{(r)}).$$

We defined the scale factor for the function F to be

$$\frac{\text{estimated error in image of } x}{\text{error in } x},$$

i.e.

$$\frac{\text{error in } x^{(r+1)}}{\text{error in } x^{(r)}}.$$

We took a guess, $x^{(1)}$ say, at the solution of $x = F(x)$, and calculated the scale factor, which we assume changes little in the neighbourhood of the solution. If the magnitude of the scale factor were greater than 1, then the error interval containing the guess and the actual solution would grow, and if the magnitude of the scale factor were less than 1, then the error interval would shrink at each application of F, and the iterative procedure would converge to the solution.

Now let us have a look at what this means in our present context. The scale factor here may be defined to be

Main Text
* * *

$$\frac{m_2(\underline{e}^{(r+1)})}{m_2(\underline{e}^{(r)})} = \frac{\text{the norm of the estimated error in } \underline{x}^{(r+1)}}{\text{the norm of the estimated error in } \underline{x}^{(r)}}.$$

Note that the scale factor cannot be negative, by our definition of a *norm*. If this scale factor is less than 1 for all r, then

$$m_2(\underline{e}^{(r+1)}) < m_2(\underline{e}^{(r)}),$$

i.e.

$$|e_1^{(r+1)}| + |e_2^{(r+1)}| + \cdots + |e_n^{(r+1)}| < |e_1^{(r)}| + |e_2^{(r)}| + \cdots + |e_n^{(r)}|.$$

This inequality is true for all r, so we can find a number k, $0 < k < 1$, such that

$$\sum_{i=1}^{n} |e_i^{(r+1)}| \leqslant k \sum_{i=1}^{n} |e_i^{(r)}|$$

$$\leqslant k^2 \sum_{i=1}^{n} |e_i^{(r-1)}|$$

$$\cdots$$

$$\leqslant k^r \sum_{i=1}^{n} |e_i^{(1)}|.$$

It follows that

$$\lim_{\substack{r \text{ large}}} \left(\sum_{i=1}^{n} |e_i^{(r+1)}| \right) \leqslant \left(\lim_{\substack{r \text{ large}}} k^r \right) \times \left(\sum_{i=1}^{n} |e_i^{(1)}| \right),$$

since $\sum_{i=1}^{n} |e_i^{(1)}|$ is a constant.

Now, since $0 < k < 1$, $\lim_{\substack{r \text{ large}}} k^r = 0$, so that

$$\lim_{\substack{r \text{ large}}} \left(\sum_{i=1}^{n} |e_i^{(r+1)}| \right) = 0.$$

So once again the scale factor holds the clue. If the scale factor is less than 1, then the sequence of norms of the error vectors converges to zero, i.e. the sequence of error vectors converges to the zero vector, and the sequence of iteration vectors $\underline{x}^{(r)}$ converges to the solution $\underline{X}$.

We now have a criterion for convergence, viz,

$$\frac{m_2(\underline{e}^{(r+1)})}{m_2(\underline{e}^{(r)})} < 1.$$

We know that

$$\underline{e}^{(r+1)} = G\underline{e}^{(r)},$$

so that this condition becomes

$$\frac{m_2(G\underline{e}^{(r)})}{m_2(\underline{e}^{(r)})} < 1,$$

and it looks as if the clue to convergence is held by G. The information contained in G will relate the norms of successive error vectors in some way.

We shall illustrate this in the two-dimensional case.

If

$$m_2(\underline{e}^{(r+1)}) \leqslant k m_2(\underline{e}^{(r)}), \quad \text{where } 0 < k < 1,$$

then the lengths of the sides of the squares are decreasing by a factor k at each step, and the arrow representing the error vector, which is trapped inside the appropriate square, is approaching the zero vector, i.e. the sequence of vectors is converging. (See Exercise 2.)

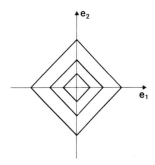

We shall now examine what the scale factor condition means in terms of the elements of G. We shall do the analysis in two dimensions for simplicity, but the arguments can easily be generalized.

Suppose

$$G = \begin{pmatrix} g_{11} & g_{12} \\ g_{21} & g_{22} \end{pmatrix}.$$

Successive error vectors are then

$$\begin{pmatrix} e_1^{(r+1)} \\ e_2^{(r+1)} \end{pmatrix} = \begin{pmatrix} g_{11} & g_{12} \\ g_{21} & g_{22} \end{pmatrix} \begin{pmatrix} e_1^{(r)} \\ e_2^{(r)} \end{pmatrix},$$

so

$$e_1^{(r+1)} = g_{11}e_1^{(r)} + g_{12}e_2^{(r)}$$

and

$$e_2^{(r+1)} = g_{21}e_1^{(r)} + g_{22}e_2^{(r)}.$$

Thus

$$m_2(\underline{e}^{(r+1)}) = |e_1^{(r+1)}| + |e_2^{(r+1)}|$$
$$= |g_{11}e_1^{(r)} + g_{12}e_2^{(r)}| + |g_{21}e_1^{(r)} + g_{22}e_2^{(r)}|$$
$$\leqslant |g_{11}e_1^{(r)}| + |g_{12}e_2^{(r)}| + |g_{21}e_1^{(r)}| + |g_{22}e_2^{(r)}|,$$

using the triangle inequality (see *Unit 22, Linear Algebra I*, section 22.1.3). Since, for any real numbers α, β,

$$|\alpha \times \beta| = |\alpha| \times |\beta|,$$

we have

$$|e_1^{(r+1)}| + |e_2^{(r+1)}| \leqslant |g_{11}||e_1^{(r)}| + |g_{12}||e_2^{(r)}| + |g_{21}||e_1^{(r)}| + |g_{22}||e_2^{(r)}|,$$

and finally

$$|e_1^{(r+1)}| + |e_2^{(r+1)}| \leqslant (|g_{11}| + |g_{21}|)|e_1^{(r)}| + (|g_{12}| + |g_{22}|)|e_2^{(r)}|.$$

Therefore if we can find a number k in the range $0 < k < 1$ such that both

$$|g_{11}| + |g_{21}| \leqslant k$$

and

$$|g_{12}| + |g_{22}| \leqslant k,$$

the scale factor condition is satisfied, and the sequence of error vectors converges to the zero vector. Thus we can test the matrix G by adding the moduli of the elements in each column. If the largest column sum is less than 1, the sequence of error vectors converges. For example, if

$$G = \begin{pmatrix} 0.3 & 0.5 \\ 0.2 & 0.1 \end{pmatrix},$$

then we have

$$|g_{11}| + |g_{21}| = 0.5 \quad \text{and} \quad |g_{12}| + |g_{22}| = 0.6.$$

If we take $k = 0.6$ (or any number between 0.6 and 1), then the scale factor condition is satisfied, so the iterative sequence produced by the matrix G converges.

(This number k, representing the maximum sum of the moduli of the elements of a column of a matrix, can itself be regarded as a norm of a matrix. In other words, it is a method of "measuring" or attaching a number to a matrix.)

Exercise 3

Exercise 3
(3 minutes)

Test each of the following matrices to see if the sequence of error vectors obtained from the equation

$$\underline{e}^{(r+1)} = G\underline{e}^{(r)}$$

is certain to converge to the zero vector. If it is, give a suitable value for the constant k.

(i) $G = \begin{pmatrix} 0.3 & 0.75 \\ 0.6 & 0.2 \end{pmatrix}$

(ii) $G = \begin{pmatrix} 0.7 & 0.05 \\ -0.5 & 0.1 \end{pmatrix}$

(iii) $G = \begin{pmatrix} 0.4 & 0.3 \\ 0.6 & 0.7 \end{pmatrix}$

(iv) $G = \begin{pmatrix} 0.33 & -0.72 \\ -0.66 & -0.27 \end{pmatrix}$ ■

Exercise 4

Exercise 4
(3 minutes)

Rearrange the following equations in such a way that the resulting iterative method converges. Hence solve the equations by iteration.

$$5x_1 + 3x_2 = 7$$
$$2x_1 - 4x_2 = 3.$$ ■

The scale factor test can be applied to a sequence of error vectors in a vector space of any finite dimension. We sum the moduli of the elements of each column of a matrix, and if all the sums are less than or equal to k, which is a positive number <1, then we have guaranteed convergence. For example, for three dimensions, we have another geometric interpretation. Using the norm $|e_1| + |e_2| + |e_3|$, we confine the arrow representing the error vector inside a box in the shape of an octahedron which gets smaller at each step if the method converges.

Discussion

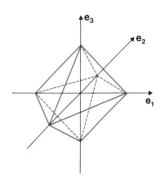

For more than three dimensions we have no geometric picture, but the algebra is just the same.

Finally, we turn to a point made in the introduction to this section. We said that the iterative method is useful when the matrix of coefficients is a sparse matrix. This is so because when there are only a few non-zero elements in each row, we can readily isolate one of the elements of the solution vector and express it in terms of relatively few elements on the right-hand side. (There may be some manipulation required to obtain such an equation for *each* element.) For example, the arrangement

$$x_1 = 0.2x_2 + 0.5x_5 + 1$$
$$x_2 = 0.4x_1 + 0.1x_3 + 2$$
$$x_3 = 0.2x_2 + 0.3x_4 + 3$$
$$x_4 = 0.5x_3 + 0.2x_5 + 4$$
$$x_5 = 0.4x_1 + 0.3x_2 + 5$$

has 3 fewer multiplications on the right-hand side of each equation than it might have. This is because the matrix of coefficients written in the form

$$\begin{pmatrix} 1 & -0.2 & 0 & 0 & -0.5 \\ -0.4 & 1 & -0.1 & 0 & 0 \\ 0 & -0.2 & 1 & -0.3 & 0 \\ 0 & 0 & -0.5 & 1 & -0.2 \\ -0.4 & -0.3 & 0 & 0 & 1 \end{pmatrix}$$

is relatively sparse. The fact that the number of multiplications required is drastically reduced would be even more pronounced if, for example, there were only 5 non-zero elements in each row of a 100×100 matrix. The final judgment on efficiency, compared with a direct method, is more difficult however, for we also have to decide how fast the iterative process converges, assuming it does converge; that is how many steps are required to obtain the required accuracy.

31

Solution 3   **Solution 3**

(i) The sequence converges. Any k such that $0.95 \leqslant k < 1$ is suitable.
(ii) The sequence does not necessarily converge.
(iii) The sequence does not necessarily converge.
(iv) The sequence converges. Any k such that $0.99 \leqslant k < 1$ is suitable.

Note that we have only shown that

condition satisfied $\Rightarrow$ convergence

and not that

condition not satisfied $\Rightarrow$ divergence;

we may still have convergence in the latter case, although it may be more difficult to prove. Thus you might like to look at the sequence generated by the G's in cases (ii) and (iii) starting with $\begin{pmatrix} 1 \\ 1 \end{pmatrix}$.   ■

Solution 4   **Solution 4**

One such rearrangement is

$$x_1 = 1.4 - 0.6x_2$$
$$x_2 = -0.75 + 0.5x_1.$$

With this rearrangement,

$$G = \begin{pmatrix} 0 & -0.6 \\ 0.5 & 0 \end{pmatrix}$$

and the iteration formula takes the form

$$\underline{x}^{(r+1)} = \begin{pmatrix} 0 & -0.6 \\ 0.5 & 0 \end{pmatrix} \underline{x}^{(r)} + \begin{pmatrix} 1.4 \\ -0.75 \end{pmatrix}.$$

Starting with any $\underline{x}^{(1)}$, we shall eventually get as close as we please to the actual solution $\begin{pmatrix} \dfrac{37}{26} \\ \dfrac{-1}{26} \end{pmatrix} \simeq \begin{pmatrix} 1.42 \\ -0.04 \end{pmatrix}.$ ■

Exercise 5 **Exercise 5**

The following is a simple library program, called $ ITER which you can use to get solutions of simultaneous equations by iteration. You ought to be able to follow how it works by inspection.

PROGRAM	COMMENT
5 PRINT "X = AX + B, A AN N * N MATRIX"	Sets the scene.
10 PRINT "N ="	
15 INPUT N	Asks for matrix size
16 IF N < = 10 THEN 20	which must be less
17 PRINT "N TOO LARGE GIVE A NUMBER FROM 1 TO 10"	than 10.
18 GO TO 10	
20 PRINT "A = "	
25 FOR I = 1 TO N	Inputs matrix A
30 FOR J = 1 TO N	row by row,
35 INPUT A(I, J)	one element for each
40 NEXT J	input request.
45 NEXT I	
50 PRINT "B = "	
55 FOR I = 1 TO N	Inputs B.
60 INPUT B(I)	
65 NEXT I	
70 PRINT "INITIAL X = "	Requests initial
71 FOR I = 1 TO N	estimate X.
72 INPUT X(I)	
73 NEXT I	
75 PRINT "WHAT NEXT? REPLY WITH −1 IF FURTHER ITERATIONS ARE REQUIRED"	
76 PRINT "+1 IF A NEW MATRIX IS TO BE INSERTED AND 0 TO TERMINATE"	
80 INPUT X	
85 IF X = −1 THEN 110	
90 IF X = +1 THEN 5	
95 IF X = 0 THEN 300	
100 PRINT "REPLY MUST BE −1, +1 OR 0"	
105 GO TO 80	
110 PRINT "HOW MANY ITERATIONS?"	
115 INPUT M	Makes sure that no
120 IF M < 21 THEN 135	more than 20 iterations
125 PRINT "GIVE A NUMBER FROM 1 TO 20"	are requested.
130 GO TO 115	

PROGRAM	COMMENT
135 FOR K = 1 TO M	
140 FOR I = 1 TO N	
141 Y(I) = B(I)	
142 NEXT I	
145 FOR I = 1 TO N	
150 FOR J = 1 TO N	
155 IF A(I, J) = 0 THEN 165	Performs the iterations.
160 LET Y(I) = Y(I) + A(I, J) * X(J)	
165 NEXT J	
170 NEXT I	
175 FOR I = 1 TO N	
176 X(I) = Y(I)	
177 NEXT I	
180 NEXT K	
185 PRINT "X = "	
190 FOR I = 1 TO N	
192 PRINT X(I),	Prints results.
194 NEXT I	
196 GO TO 75	
300 END	

(i) As it stands, this program cannot tell you whether the matrix A which you input makes the iteration likely to converge. Write a set of instructions which will modify the program so that it will tell you the maximum value of $\sum_{i=1}^{n} |a_{ij}|$ for $j = 1, \ldots, n$, and refuse to iterate unless this value is less than 1.

(ii) Use the program to solve the following 5×5 system to two decimal places.

$$
\begin{aligned}
x_1 - 0.2x_2 \qquad\qquad\quad - 0.5x_5 &= 1 \\
-0.4x_1 + 1.1x_2 - 0.1x_3 \qquad\qquad &= 2 \\
- 0.2x_2 + \quad x_3 - 0.3x_4 \qquad &= 3 \\
- 0.5x_3 + 1.2x_4 - 0.2x_5 &= 4 \\
-0.4x_1 - 0.3x_2 \qquad\qquad + \quad x_5 &= 5.
\end{aligned}
$$

28.2.3   Summary

To solve the system of equations represented in matrix form by the equation

$$A\underline{x} = \underline{b},$$

we rearrange it in the form

$$\underline{x} = G\underline{x} + H\underline{b}.$$

This leads to a possible iterative sequence defined by

$$\underline{x}^{(r+1)} = G\underline{x}^{(r)} + H\underline{b} \qquad (r = 1, 2, \ldots),$$

which will converge if the sum of the moduli of the elements in each column of G is less than 1. We proved this result by examining the behaviour of the error vectors

$$\underline{e}^{(r)} = \underline{x}^{(r)} - \underline{X} \qquad (r = 1, 2, \ldots),$$

which satisfy

$$\underline{e}^{(r+1)} = G\underline{e}^{(r)} \qquad (r = 1, 2, \ldots),$$

where $\underline{X}$ is the exact solution vector.

Solution 28.2.2.5 **Solution 28.2.2.5**

(i) A suitable set of instructions would be:

  46 GOTO 200

200 LET P = 0

205 FOR J = 1 TO N

210 LET Z = 0

215 FOR I = 1 TO N

220 LET Z = Z + ABS(A(I, J))

225 NEXT I

230 IF Z < P THEN 240

235 LET P = Z

240 NEXT J

245 PRINT "MATRIX NORM ="; P

250 IF P < 1 THEN 50

255 PRINT "MATRIX NORM TOO BIG"

(ii) A suitable rearrangement of this is

$$\underline{x} = \begin{pmatrix} 0 & 0.2 & 0 & 0 & 0.5 \\ 0.4 & -0.1 & 0.1 & 0 & 0 \\ 0 & 0.2 & 0 & 0.3 & 0 \\ 0 & 0 & 0.5 & -0.2 & 0.2 \\ 0.4 & 0.3 & 0 & 0 & 0 \end{pmatrix} \underline{x} + \begin{pmatrix} 1 \\ 2 \\ 3 \\ 4 \\ 5 \end{pmatrix}.$$

The solution to two decimal places is

$$\underline{x} = \begin{pmatrix} 6.44 \\ 4.72 \\ 6.16 \\ 7.40 \\ 8.99 \end{pmatrix}.$$

∎

28.3 ILL-CONDITIONED SYSTEMS OF EQUATIONS

In this section we look at a particularly disastrous way in which errors can sometimes accumulate when we solve simultaneous equations based on inexact data or when we introduce rounding errors during solution. In fact, the error accumulation may even render the results of solving the simultaneous equations virtually useless. When small changes in the data have a large effect on the solution of a system of equations, then we say that the system is ill-conditioned. This term has no precise definition but is used in the same relative sense that adjectives like "small" are used in English. "Small changes in the data" producing "large changes in the result" is simply one way of describing the phenomenon of ill-conditioning; the adjectives used indicate the imprecision of the concept. A numerical illustration of ill-conditioning is given in the following example.

Example 1

Example 1

In a reconstruction of a crime, bullet holes centred at *A* and *B* were found in a double-glazed window at distances apart indicated on the diagram, the measurements being taken to the centres of the holes. All the measurements were taken *to the nearest centimetre*. Other evidence showed that the gun was at the level *CD* when fired. How accurately can we determine the position from which the gun was fired?

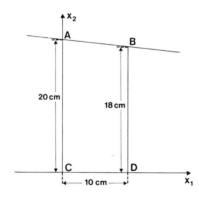

Before following through the solution to this example, you may find it helpful to assume that the measurement of 10 cm is *exact* and draw accurately, on a piece of graph paper, the two most extreme possible trajectories of the bullet, assuming them to be straight lines.

Solution of Example 1

If we fit x_1 and x_2 co-ordinate axes on to the diagram, we are effectively finding the intersection of the straight line *AB* with the x_1-axis *CD*.

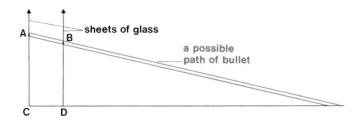

37

The equation of the straight line AB is

$$x_2 = mx_1 + c,$$

(See RB3)

where (in theory) we can find m and c from the data. The distance x_G from C at which the bullet was fired is the value of x_1 when $x_2 = 0$, i.e.

$$x_G = -\frac{c}{m}.$$

If we assume the data to be exact, then we know that AB passes through the points $(0, 20)$ and $(10, 18)$, so that

$$20 = c$$

$$18 = 10m + c,$$

whence $m = -0.2$. Hence $x_G = \dfrac{20}{0.2}\,\text{cm} = 100\,\text{cm} = 1\,\text{m}$.

But the data are *not* exact: c could be 19.5 and then m could be as *small* as the value given by the equation

$$18.5 = 10.5m + 19.5,$$

i.e.

$$m = \frac{-1}{10.5};$$

so that x_G could be as *large* as $19.5 \times 10.5\,\text{cm} = 204.75\,\text{cm}$. That is, the approximate value for x_G (1 m) could be over 1 m in error. In fact, all we can say is that

$$x_G \in [64.92, 204.75]\,\text{cm}.$$

In other words, the distance the gun was fired from the window could be anything between just over 0.5 m and just over 2 m. This is not a very accurate result when you consider the apparent accuracy of the original measurements. ■

In geometric terms, in the last example we were trying to find the intersection of a pair of nearly parallel straight lines (the x_1-axis was one of these). If there is a small error in the original data, it can result in a considerable error in the solution. One possible way of envisaging what is happening is to imagine two torches, or searchlights, with their beams crossed.

Discussion

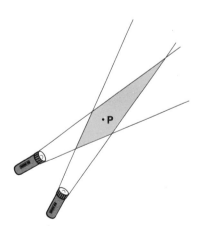

The exact solution, if the data are exact, is represented by a point P. With inaccuracies in the data, represented by the diverging beams from the torches, the straight lines could be anywhere within the beams. The point representing the true solution could be anywhere within the "area of intersection" which is shaded in the diagram.

Intuitively, the more nearly parallel to each other the axes of the torch beams become, the larger the shaded area, and the further away the true solution can be from the approximate solution.

Now let us look at a more general case in two dimensions in the following exercise.

Exercise 1

Solve the following simultaneous equations for x_1 and x_2:

$$x_1 + \quad\quad x_2 = 1$$
$$x_1 + (1 + \varepsilon)x_2 = 2,$$

where ε is some real number, by the Gauss elimination method. Write down the solutions corresponding to

(i) $\varepsilon = 0.01, 0.02$
(ii) $\varepsilon = 2.01, 2.04$

respectively. Compare the percentage changes in the *coefficient* of x_2 in the second equation for cases (i) and (ii) with the percentage changes in the corresponding *solution* for x_2. ∎

This last exercise was rather hypothetical. However, the rounding errors which occur both in the initial data from a practical problem and in the solution process itself are not hypothetical. These small rounding errors can readily lead, in ill-conditioned equations, to highly inaccurate results. A simple way to test for such ill-conditioning is to do what we did in the last exercise, that is, to make a small change in some coefficients to see what effects there are on the solution, but this is difficult to do with larger sets of equations, particularly when a solution set looks acceptable in the physical sense. There are more sophisticated methods which give a practical indication when ill-conditioning is present, but most of these will involve us in too many technicalities. We shall describe just one such method.

Theoretically, ill-conditioning can be described in the following way. Given a system of simultaneous equations in matrix form

$$A\underline{x} = \underline{b},$$

the system is ill-conditioned if the n columns of the matrix are *almost* linearly dependent (note the vagueness again), or, in other words, if the matrix is *nearly singular*. This means that all the elements of A are close to the corresponding elements of some singular matrix. In the last exercise, for example, A would be

$$\begin{pmatrix} 1 & 1 \\ 1 & 1 + \varepsilon \end{pmatrix}.$$

Clearly if ε were small (we found this made the equations ill-conditioned), a small change, that is a change of $-\varepsilon$ in a_{22}, would turn the matrix into the singular matrix

$$\begin{pmatrix} 1 & 1 \\ 1 & 1 \end{pmatrix}$$

in which the two columns are obviously linearly dependent.

Exercise 2

Determine the inverse matrix A^{-1} of

$$A = \begin{pmatrix} 1 & 1 \\ 1 & 1 + \varepsilon \end{pmatrix}.$$

What are the elements of A^{-1} if $\varepsilon = 0.01$? ∎

Exercise 1
(4 minutes)

Discussion
* *

Exercise 2
(4 minutes)

(*continued on page 40*)

Solution 1 **Solution 1**

$$x_1 + \quad x_2 = 1$$
$$x_1 + (1 + \varepsilon)x_2 = 2$$

Carrying out the Gauss elimination method gives

$$x_1 + x_2 = 1$$
$$\varepsilon x_2 = 1,$$

which gives

$$x_2 = \frac{1}{\varepsilon} \quad (\varepsilon \neq 0),$$

and from back-substitution,

$$x_1 = 1 - \frac{1}{\varepsilon}$$

	(i)		(ii)	
ε	0.01	0.02	2.01	2.04
Solution to 2 decimal places	$(-99, 100)$	$(-49, 50)$	$(0.50, 0.50)$	$(0.51, 0.49)$

Percentage change of coefficient in x_2 is 1% to one place of decimals in each case.

In (i) the solution x_2 changes by 50%; in (ii) by 2%. This indicates that "when ε is small the equations are ill-conditioned" means in this case "small compared with unity". ∎

Solution 2 **Solution 2**

$$A^{-1} = \begin{pmatrix} 1 + \dfrac{1}{\varepsilon} & -\dfrac{1}{\varepsilon} \\ -\dfrac{1}{\varepsilon} & \dfrac{1}{\varepsilon} \end{pmatrix}$$

$$\begin{pmatrix} 101 & -100 \\ -100 & 100 \end{pmatrix}$$ ∎

(continued from page 39)

This last exercise shows one feature of the inverse of the matrix of coefficients of an ill-conditioned system of equations; that is, its elements are relatively large when compared with the elements of the original matrix. This leads to one final point about systems of equations which may be ill-conditioned. A recommended and useful means of checking the accuracy of an estimated solution $\underline{x}_s$ of

$$A\underline{x} = \underline{b}$$

is to premultiply $\underline{x}_s$ by A and obtain a vector $\underline{b}_s$. That is,

$$A\underline{x}_s = \underline{b}_s.$$

If the elements of $\underline{b}_s$ differ very little from the elements of $\underline{b}$, then we would normally assume that the solution is fairly accurate. (Remember that

Discussion
* *

there will almost always be some inaccuracies from the rounding errors in a real computation of any length.) If the system of equations is ill-conditioned, however, the solutions can still be grossly in error; for if we subtract the two equations above we get

$$A(\underline{x}_s - \underline{x}) = \underline{b}_s - \underline{b},$$

or, using the error vector notation,

$$A\underline{e} = \underline{E}$$

where

$$\underline{e} = \underline{x}_s - \underline{x} \quad \text{and} \quad \underline{E} = \underline{b}_s - \underline{b}.$$

Then we shall have

$$\underline{e} = A^{-1}\underline{E}.$$

If A^{-1} contains large elements, it is intuitively obvious that $\underline{e}$ can be large even though $\underline{E}$ is small. So, whenever we suspect ill-conditioning, it is worth having a look at the size of the elements of A^{-1} relative to the elements of A.

Exercise 3

Exercise 3
(4 minutes)

If

$$A = \begin{pmatrix} 1 & 1 \\ 1 & 1.01 \end{pmatrix},$$

find an $\underline{E}$ with norm (sum of the moduli of the elements) <0.01 which leads to an $\underline{e}$ with norm >1. ■

Finally, there is the problem of what to do about ill-conditioning when it occurs and we do recognize it. It may well be that reformulation of the problem in terms of different variables, as may be possible with sets of ill-conditioned simultaneous equations arising from problems involving engineering structures, will lead to a well-conditioned set of equations. Otherwise we can use double precision arithmetic in the calculation, that is, carry twice as many digits throughout the work, in an attempt to get better results, but if the equations are badly ill-conditioned this will still be of no avail. Then the advice is — forget the problem, or quote the result to whatever accuracy is obtained, however bad.

Discussion

Summary

When the matrix of coefficients is nearly singular, that is, a small change in some elements of the matrix will produce a singular matrix, the matrix equation

$$A\underline{x} = \underline{b}$$

Summary

is said to be ill-conditioned. This usually means that the solution is highly unreliable, particularly if the original data, from which the matrix equation arose, were inexact.

Solution 3

Solution 3

One example is

$$\underline{E} = \begin{pmatrix} -0.004 \\ 0.004 \end{pmatrix}$$

Then $m_2(\underline{E}) = 0.008 < 0.01$ and

$$\underline{e} = \begin{pmatrix} 101 & -100 \\ -100 & 100 \end{pmatrix} \begin{pmatrix} -0.004 \\ 0.004 \end{pmatrix} = \begin{pmatrix} -0.804 \\ 0.8 \end{pmatrix}$$

with $m_2(\underline{e}) = 1.604$. ∎

28.4 CONCLUSION

In this text we have examined the practical problem of how systems of simultaneous linear equations, particularly large systems, can be solved.

In section 28.3 we drew attention to the phenomenon of ill-conditioning which can arise in such systems of equations, and which may make it very difficult, or even impossible, to find any reliable or useful solution to the equations.

We have demonstrated the basic direct and indirect methods of solution. Many refinements grow out of these and can be found, for example, in Fox, *Numerical Linear Algebra* (see Bibliography). All we have done here is to examine the basic methods and to look at their efficiency and accuracy.

Postscript

"That's carrying things a step too far, I draw the line at that."

Harry B. Smith,
We Draw the Line at That

Unit No.		Title of Text
1		Functions
2		Errors and Accuracy
3		Operations and Morphisms
4		Finite Differences
5	NO TEXT	
6		Inequalities
7		Sequences and Limits I
8		Computing I
9		Integration I
10	NO TEXT	
11		Logic I — Boolean Algebra
12		Differentiation I
13		Integration II
14		Sequences and Limits II
15		Differentiation II
16		Probability and Statistics I
17		Logic II — Proof
18		Probability and Statistics II
19		Relations
20		Computing II
21		Probability and Statistics III
22		Linear Algebra I
23		Linear Algebra II
24		Differential Equations I
25	NO TEXT	
26		Linear Algebra III
27		Complex Numbers I
28		Linear Algebra IV
29		Complex Numbers II
30		Groups I
31		Differential Equations II
32	NO TEXT	
33		Groups II
34		Number Systems
35		Topology
36		Mathematical Structures